Fiber Optic Communications For Beginners
-The Basics

Eric R. Pearson

CFOS/T/C/S/I

Pearson Technologies Inc.
4671 Hickory Bend Drive
Acworth, GA 30102-6340
770-490-9991
www.ptnowire.com
fiberguru@ptnowire.com

36 Years Of Superior Fiber Optic Training And Consulting

File: Fiber Optic Communications For Beginners, v0
ISBN 978-1517789022
12 October 2015

© Pearson Technologies Inc.

Table Of Contents

1 Introduction 6
1.1 10 Advantages 6
1.1.1 Unlimited Bandwidth 6
1.1.2 Transmission Distance 7
1.1.3 EMI And RFI 7
1.1.4 Low Bit Cost 7
1.1.5 Dielectric Construction 8
1.1.6 Small Size 8
1.1.7 Light Weight 9
1.1.8 Installation Ease 9
1.1.9 Security 9
1.1.10 Power Requirements 10
1.2 Fiber Networks 10
1.3 Link Description 10
1.4 Full Duplex Links 11
1.4.1 Pulse Changes In Link 11

2 Signal Types 13

3 Optoelectronics 14
3.1 Introduction 14
3.2 Pulse Changes 14
3.3 3 Source Types 15
3.4 Multiple Wavelengths 16

4 Passive devices 17
4.1 Introduction 17
4.2 Types 17
4.2.1 Couplers 17
4.2.2 Splitters 17
4.2.3 Optical Amplifiers 18
4.2.4 Dispersion Compensators 19

| | | 4.2.5 | Switches And Routers | 19 |

5 Fiber 21

    5.1 Fiber Function 21

    5.2 Fiber Structure 21

    5.3 Fiber Types 21

        5.3.1 Multimode 22

        5.3.2 Singlemode 22

        5.3.3 Bend Sensitivity 23

6 Cables 24

    6.1 Introduction & Structure 24

    6.2 Cable Designs 24

        6.2.1 4 Loose Tube Designs 24

        6.2.2 2 Tight Tube Designs 26

    6.3 Performance Requirements 26

    6.4 Color Codes 26

7 Connectors 28

    7.1 Functions 28

    7.2 Connector Types 28

    7.3 Color Coding 30

    7.4 Installation Methods 31

        7.4.1 Polish Methods 31

        7.4.2 No Polish Methods 31

8 Splices 33

    8.1 Introduction 33

    8.2 Two Methods 33

    8.3 Splicing Steps 33

9 Testing 35

    9.1 Types 35

    9.2 Insertion Loss Testing 35

    9.3 OTDR Testing 36

    9.4 Reflectance Testing 37

9.5 Protocol Testing 38
10 Design Concerns 39
11 About The Author 41

1 INTRODUCTION

This is an introductory text for those interested in fiber optic communications. This text provides a framework on which the student can organize additional, detailed knowledge. It is not designed to be comprehensive. For a comprehensive understanding, see these training texts:

> Professional Fiber Optic Installation, v.9- The Essentials
>
> Mastering The OTDR-Trace Acquisition And Analysis
>
> Mastering Fiber Optic Network Design- The Essentials
>
> Mastering Fiber Optic Connector Installation

Single copies of these texts are available from Amazon.com. Multiple copies of these texts are available from Pearson Technologies Inc.

The words in **bold print** are the important technical terms. Recognition of these terms is essential to understanding the subtleties of this powerful and exciting technology.

This text is a result of this author's 38 years in fiber optic communications. During this time, this author has trained more than 8800 people in more than 530 presentations. This experience has shown this author the concepts that people understand easily. These are the concepts in this text. Enjoy.

1.1 10 ADVANTAGES

In general, optical fiber is used in two situations. The first situation is that in which its capabilities exceed the limitations of other communication media. The second is those situations in which fiber transmission provides a combination of advantages, often including that of reduced cost.

Optical fiber is the medium of choice when its one of it's ten characteristics favor its use. These characteristics are:

- ➢ Nearly unlimited bandwidth
- ➢ Long transmission distance
- ➢ Low cost per bit
- ➢ EMI and RFI immunity
- ➢ Dielectric construction
- ➢ Small size
- ➢ Light weight
- ➢ Ease of installation
- ➢ Intrinsically secure transmission
- ➢ Reduced power requirements

1.1.1 UNLIMITED BANDWIDTH

Optical fiber has essentially unlimited bandwidth. While 'unlimited bandwidth' may sound like an exaggeration, it is a realistic and reasonable description of fiber capacity. In 2000, a study by Lucent Technologies indicated the theoretical capacity of one singlemode fiber is on the order of 200 Tbps, or 200 million Mbps. Such a capacity deserves the term 'essentially unlimited'. With the advent of the latest technology, coherent transmission, it is possible that this limit will increase by a factor of eight.

One part of this high capacity results from the ability to modulate light sources at very high data rates. At the time of this writing, a practical maximum, single wavelength data rate is 40 Gbps. This rate is 4,000 times the rate of the early, Ethernet fiber optic transceivers!

A second part of high capacity results from the ability to transmit multiple wavelengths on one fiber simultaneously. When properly implemented, multiple wavelengths do not interfere with each other.

Such transmission has three designations:

- **Wavelength division multiplexing (WDM)**
- **Coarse wave-length division multiplexing (CWDM)**
- **Dense wavelength division multiplexing (DWDM)**.

A third part of high capacity results from the ability of an optical pulse to transmit more than one data bit. At the time of this writing, this technology, **coherent transmission**, enables transmission of eight electrical bits with one optical pulse.

This 'essentially unlimited' bandwidth means that a simple replacement of the optoelectronics on the fiber ends increases link bandwidth. This upgradeability results in low life cycle cost and low total cost of ownership.

1.1.2 TRANSMISSION DISTANCE

Optical fiber allows extremely long transmission distance without return to the electrical regime. Long transmission distance results in a reduction in the cost of mid-span signal repeaters and regenerators. Elimination of these electronics reduces hardware and maintenance costs significantly.

While not a current 'champion' result, Williams Communications demonstrated the ability to transmit 5000 km (3100 miles) in the optical regime. 5000 km is approximately the distance between Boston MA and Los Angeles CA!

In the early 1990s, long distance capability provided two major benefits to the CATV industry. This long transmission distance enabled a major cost reduction through a reduction in the number of satellite 'farms' necessary to support a service area. In addition, long transmission distance enabled CATV companies to reduce the number of coax amplifiers between a satellite down link and set top boxes. This reduction resulted in reduced equipment and maintenance costs, and improved signal quality.

1.1.3 EMI AND RFI

Long transmission distance capability results primarily from low power loss (attenuation rate) and low pulse dispersion. However, a third property is equally important.

This property, immunity to electromagnetic interference (EMI) and radio frequency interference (RFI), enables optical signals to travel long distances without the need for signal correction due to interference from EM and RF signals in the environment.

1.1.4 LOW BIT COST

The combination of multiple wavelength transmission, low power loss, low pulse dispersion, and EM and RF immunity results in low cost per bit. This low cost has made fiber the medium of choice for long distance, high capacity communication.

Consider an example of a telephone line. The installed fiber cable cost for a 64kbps voice channel is approximately

$0.0000000000000025/voice line/mile.

A data network line operating at 1 Gbps is much more expensive:

$$0.0000000000016/1 Gbps line/mile

This low cost has resulted in the displacement of satellites as the 'king' of long distance communication. Now satellites are a back up for optical fiber transmission, a reverse of the original relationship!

At the time of this writing, some local area networks (LANs) with centralized backbones have a total initial installed cost that is lower than that of traditional horizontal UTP, vertical fiber networks. Such networks are known as **fiber to the desk (FTTD)** and **collapsed backbone** networks.[1]

This cost advantage results from a reduction in the costs of telecommunication rooms and switches. Use of fiber significantly reduces the cost of such rooms and the cost of support for such rooms.[2]

---

[1] See the cost model offered by the Fiber Optic LAN Section (FOLS) of the TIA at www.fols.org. The FOLS and Pearson Technologies co-developed this model.

[2] See the fiber to the desk cost comparison at www.fols.org. Pearson Technologies Inc. and the FOLS developed this cost comparison.

Finally, the low cost of optical fiber has led to increased implementation of fiber-to-the-home (FTTH) networks. The characteristics of high bandwidth and multiple wavelength transmission enable cost effective delivery of voice, video, and Internet services, also known as '**triple play**' services.

1.1.5 DIELECTRIC CONSTRUCTION

Conductive cables must be grounded and bonded to prevent induced currents from entering a building. Such currents are caused by lightning and ground potential rise. Such currents can injure people and damage electronics.

Optical fiber cables can be made without conductive elements. Such **dielectric** construction eliminates both the initial installed cost and the maintenance cost of grounds and bonds. After all, grounds and bonds do not last forever.

1.1.6 SMALL SIZE

The small size of optical fibers and their cables results in reduced system cost. For example, large cities with filled underground conduit systems have two methods of increasing telephone capacity: 1) dig up the streets to install more conduits or 2) replace large copper cables with small fiber cables.

This author's rough estimate of the replacement ratio is 12,500, 3" diameter, 900 pair cables to one 0.5" diameter fiber cable with one wavelength transmission. With dense wavelength division multiplexing (DWDM) allowing 200 wavelengths per fiber, this replacement ratio

2,500,000! The cost advantage of using fiber instead of digging up streets is significant.

1.1.7 LIGHT WEIGHT

Optical fiber cables are significantly lighter than copper cables. As a result, fiber finds use in field tactical military, shipboard, and aircraft applications.

In field tactical applications, reduced weight enables soldiers to carry increased cable lengths. Increased lengths enable placement of electronic monitoring equipment at the front line while the monitoring personnel are in a safe location away from that line. In addition, the non-radiating nature of optical fibers prevents the enemy from detecting the equipment location.

In shipboard applications, the light weight of optical cables increases the stability of the ships by reducing the weight above the waterline. Finally, in aircraft applications, the light weight increases mission endurance.

1.1.8 INSTALLATION EASE

Because of their light weight and small size, fiber cables are easy to install. In addition, fiber connector installation methods have advanced sufficiently to enable installation by junior and senior high school students with minimal training! In the mid 1990s, the author's sons demonstrated this ability.

1.1.9 SECURITY

Fiber transmission is nearly completely secure. Its security comes from two aspects.

- No radiation
- Difficulty in tapping signal without such tapping being detected

An optical signal travels in the center of the fiber, known as the core or the mode field diameter. As such, no signal is radiated externally to the fiber or its cable.

In addition, any effort to tap optical power from the fiber reduces the power delivered to the receiver. With the addition of a simple signal intensity monitoring circuit to the receiver, a transmission system becomes secure.

Of course, the desire for security can be expanded beyond these two aspects. For example, a fiber transmission system developed in the 1980s required six fibers. The system monitored the signal intensity on all fibers. If the signal level dropped by as little as 0.2 dB, the system sounded alarms.

Two active fibers carried the encoded signal. The other fibers carried garbage signals. Periodically, the signal switched from the active fibers to those that carried the garbage signals. Any attempt to capture and decode the signal would be useless, as there would be no discernable difference between the signal and the garbage. Other than the fact that the signal could not be decoded.

Some DWDM telephone links use a approach, similar to that described in the last paragraph, to monitor link status. One of the wavelengths, usually 1625 nm, is an OTDR signal.

The OTDR monitors the loss along the entire link continuously. When an increase in the power loss occurs anywhere along the link, the monitoring equipment activates an alarm and indicates the location.

1.1.10 POWER REQUIREMENTS

The devices that convert the electrical signal to an optical signal are 'optoelectronics'. Optoelectronics require less power than do electrical transmitters and receivers in routers and switches.

Reduced power results in two benefits. The first benefit is the direct reduction in power. The second benefit is a reduction in power for cooling network electronics. This reduction results from the simple relationship that reduced power results in reduced heat generation.

1.2 FIBER NETWORKS

In the most general sense, a network is a series of electronic devices that communicate **digital** information with each other.

The network can be:

- A local area network (**LAN**)
- A data network of LANs
- A storage area network (**SAN**)
- A wide area network (**WAN**)
- A long distance, or long haul, or telephone network
- A **FTTH** network
- A passive optical LAN (**POLAN**)
- A CATV network
- A process control network

This communication is via connections between electronic devices. A point-to-point **link** forms this connection. 'Point-to-point' means that electronic conversion and distribution of the electrical signals occur only at the link ends.

> From the installer's point of view, **link** is the most important term, as all installation actions are performed on the components of each of the links.

> Incorrect installation procedures reduce the optical power delivered to the receiver on a link and may reduce the reliability of the components of a link. Thus, **power loss is the installer's most important concern**.

1.3 LINK DESCRIPTION

The simple fiber-optic link involves single wavelength transmission. A simple fiber-optic link contains optoelectronics at the ends, connectors, fiber in cable, and splices (Figure 1-1).

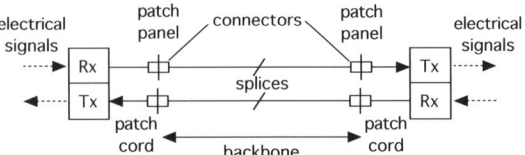

Figure 1-1: Generic Link

For descriptions of complex fiber optic transmission links, see chapters 1 and 7 of Professional Fiber Optic Installation, v.9. Such links can include multiple wavelengths, splitters, couplers, and optical amplifiers.

The transmitter and receiver are known as 'optoelectronics' because they function with both electrical and optical signals. The transmitter optoelectronics convert an electrical signal to an optical signal. There are two types of conversion.

The first type of conversion is of each electrical bit to an optical pulse. In most fiber optic data networks, this conversion occurs with one of two modifications. The first modification is conversion of four electrical bits to five optical pulses. Such conversion increases transmission accuracy. Such conversion occurs in Ethernet data networks operating at and above 100Mbps, aka Fast Ethernet.

The second modification involves scrambling the signal at the transmitter and unscrambling the signal at the receiver. Such scrambling occurs in **Synchronous Optical Networks (SONET)**, and in **passive optical networks (PONs)**.

The receiver optoelectronics perform the conversion that is the reverse of the transmitter optoelectronics.

There are two types of conversion. The first type of conversion is **direct detection**. That is, the receiver converts each optical pulse to an electrical bit.

The second type of conversion is **coherent transmission**. In coherent transmission, the receiver analyzes the optical pulse to recover the multiple electrical bits transmitted by that pulse.

1.4 FULL DUPLEX LINKS

Transceivers perform this conversion. Transceivers are optoelectronic devices that provide simultaneous optical to electrical and electrical to optical conversion; i.e., simultaneous transmission and reception. Thus, transceivers enable **full duplex** communication (Figure 1-2).

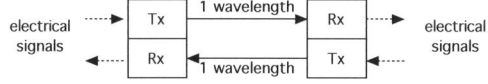

Figure 1-2: Full Duplex Transmission

In most networks, duplex transmission requires two fibers, one for transmission and one for reception. In wavelength division multiplexing, the use of two fibers is not required.

> ➢ Installers need remember that most communication requires two fibers.
> ➢ A common problem is crossed fibers.

With crossed fibers, some test results can be proper. However, communication will not occur. For the obvious reason, transmitters do not communicate properly with transmitters!

The fiber patch cables commonly used for duplex transmission are two simplex cables or a single, 'zip cord' duplex cable. Duplex cables are preferred. In contrast, UTP networks require two pairs of conductors. Thus, a fiber pair is not the same as a UTP pair.

1.4.1 PULSE CHANGES IN LINK

Each optical pulse or symbol travels through the fiber and optical connections. During such travel, the pulse or symbol experiences three changes.

The first change is a reduction in signal strength. This reduction occurs in the fiber and connections. In this text, we call the strength reduction in fiber '**attenuation**,' and that in connections, '**loss**'. When properly installed, connections cause low power loss. Low power loss results from precise alignment of the cores of the connected fibers (Figure 1-3).

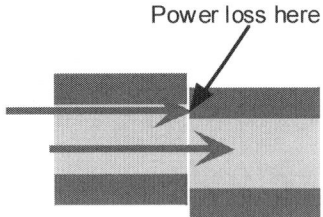

Figure 1-3: Connector Power Loss Due to Lateral Offset

The second change is a widening or spreading of the pulse width. Such spreading is known as '**dispersion**'. We address dispersion in Chapter 3.

The third change can occur in connections. Connections may create **reflections**. These reflections were called '**reflectance**'. The latest data standards use the term '**back reflection**'. This reflection appears in the OTDR trace, which we present in Testing.

2 SIGNAL TYPES

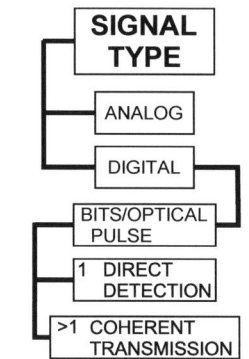

Figure 2-1: Signal Types

The signal carried on the fiber is either **analog** or **digital**. An analog signal carries the information by being modulated. This modulation can be of the **amplitude** (**AM**), or of the **frequency** (**FM**). Fiber is used to carry analog signals of cameras in some security systems.

Digital communication is used in almost all networks for communication of telephone, cable TV, and data signals. Digital communication has the prime advantage of enabling repeated amplification and regeneration without significant signal distortion. In analog communication, repetitive amplification and regeneration results in build up of signal distortion.

A second of advantage of digital communication is increased power loss allowable between transmitter and receiver. In analog transmission, this allowable power loss is significantly less that in digital transmission.

At time of this writing, digital communication is of two types: direct detection and coherent transmission. In direct detection systems, optical pulses having different power levels indicate the ones and zeros. Zeroes can be indicated by a power level of zero (**return to zero, RZ**) or a power level above zero (**no-return to zero, NRZ**).

In coherent transmission, each optical bit, known as a **symbol**, carries multiple electrical bits. This is possible through separation and modulation of the pulse height, the separate **polarizations**, and the **phase**. At the time of this writing, a single optical pulse can carry eight digital bits. Coherent transmission is possible only with high-speed digital signal processing at both the transmitter and the receiver.

3 OPTOELECTRONICS

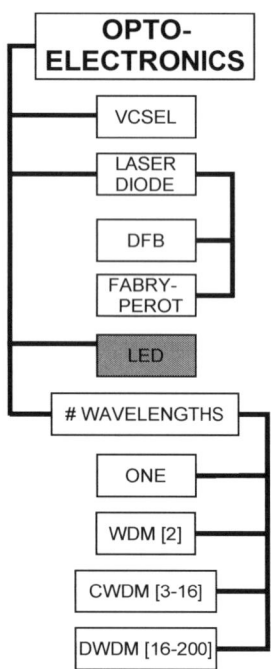

Figure 3-1: Optoelectronics

wavelengths is known as the **'spectral width'** Figure 3-2).

	Multi-	mode	Single-	mode
Data	850 780	1300	1310	
CATV			1310	1550
Telephone			1310	1550
DWDM CWDM			1310-	1550 1625
WDM	850	1300	1310	1550
FTTH/PON			1310	1490 1550

Table 3-1: Common Wavelengths

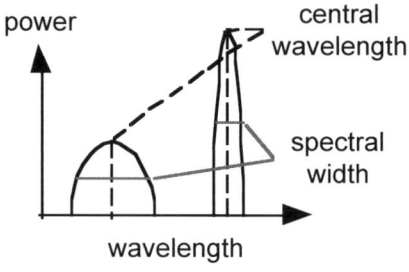

Figure 3-2: Spectral Output From Transmitters

3.1 INTRODUCTION

The optoelectronics are the devices at the ends of an optical link that perform the function of conversion. The transmitter converts an electrical bit, or multiple electrical bits, to an optical pulse or symbol. The receiver performs the reverse conversion.

The transmitter creates light of a specific color. The technical term for color is **wavelength**. Wavelength is measured in nanometers (**nm**). The wavelength of the light determines the attenuation rate and dispersion of light in the fiber. Wavelengths vary by application Table 3-1).

The term wavelength implies that the light has a single wavelength. However, the opposite is true. The optoelectronics create light in a range of wavelengths around a 'central wavelength'. The range of

Each of the wavelengths within the spectral width travels at a slightly different speed in the fiber. A consequence of these differing speeds is dispersion, which we present below.

3.2 PULSE CHANGES

Light experiences two changes as it moves through the fiber. The first change is a decrease in intensity. This decrease is called **attenuation**. Attenuation is measured in **dB/km**. The general trend is for a decrease in attenuation rate with an increase in wavelength.

The second change is a change in the width of the optical pulse. As a pulse of light moves through the fiber, the pulse becomes broader in time. This pulse broadening, or pulse spreading, is called **dispersion** (Figure 3-3).

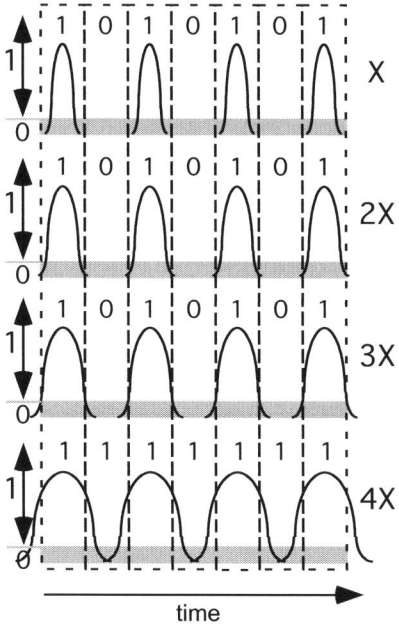

Figure 3-3: Dispersion At Multiple Transmission Distances

Because of dispersion, the pulse width at the transmitter must be less than the time interval determined by the **bit rate**, commonly, but incorrectly, called **bandwidth**. For example, a 1 Gbps data rate results in a time interval of one billionth of a second at the receiver. Since dispersion occurs, the pulse width at the transmitter must be less than 1 billionth of a second.

The general trend is for increasing dispersion on both sides of an optimum wavelength. This optimum wavelength is that at which the fiber exhibits its minimum dispersion. This optimum wavelength is known as the **zero dispersion wavelength (ZDW)** (Figure 3-4).

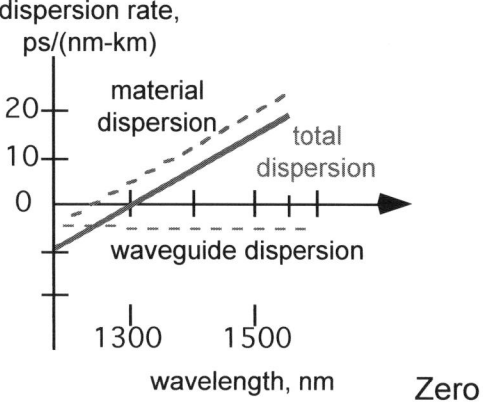

Figure 3-4: Zero Dispersion Wavelength Of G.652 Fiber

3.3   3 SOURCE TYPES

The optoelectronics convert the electrical signal into an optical signal with a light source. Three types of light sources are in use:

- a vertical cavity, surface-emitting laser (**VCSEL**),
- a laser diode (**LD**),
- light emitting diode (**LED**).

The **vertical cavity, surface emitting laser** is used with multi-mode fibers with core diameters of 50 μ to transmit signals at either one or 10 Gbps. Industry literature suggest that the current maximum rate is 25 Gbps.

Laser diodes are used with single-mode fibers at wavelengths of 1310 nm, 1550 nm, and at wavelengths between these two values. Laser diodes are directly modulation at bit rates up to 25 Gbps. Industry literature suggest that the current maximum rate is 40 Gbps.

There are two types of laser diodes: **distributed feedback (DFB)** and a **Fábry Perot**. These two sources differ in four characteristics. The first characteristic is the manner in which they create light. The second characteristic is the difference in their spectral widths. A DFB laser has a spectral width narrower than (i.e., superior to) that of the Fábry Perot laser. The third characteristic is the stability of the wavelength, with the DFB laser being more stable than the other. The fourth characteristic is the linearity of the curve of the conversion of the drive current to light output. Again, the DFB laser has superior performance.

A light emitting diode is used on multimode fibers at wavelengths of 850 and 1300 nm. LEDs are used at relatively low bit rates, roughly 200 Mbps and below. Because of this obvious bandwidth limitation, LEDs are rarely used in new networks.

3.4 MULTIPLE WAVELENGTHS

A single set of optoelectronics creates an optical signal within a spectral width around a central wavelength. However, multiple central wavelengths can travel on the same fiber without interfering with one another. This wonderfully advantageous property of light enables optical fiber to transmit extremely high bandwidths. Simultaneous transmission of multiple wavelengths on the same fiber is known as **wavelength division multi-plexing**.

Wavelength division multiplexing is of three types (Figure 3-5). **WDM** refers to simultaneous transmission of two widely separated wavelengths on the same fiber. For multimode fibers, widely separated means **850 nm** and **1300 nm**. For singlemode fibers, widely separated means **1310 nm** and **1550 nm**.

CWDM refers to simultaneous transmission of up to 16 wavelengths with separations of 20 nm on the same fiber. Historically, CWDM has been used in metropolitan areas. However, CWDM is used in **fiber to the home networks (FTTH)** and **passive optical local area networks (POLANs)**. Both FTTH and POLANs use a single fiber for transmission and reception.

DWDM refers to simultaneous transmission of up to 200 wavelengths with separations of as little as 0.4 nm on the same fiber (G.692). While the G.692 standard allows 200 wavelengths, Lucent Technologies demonstrated the ability to transmit 1000 wavelengths. Historically, DWDM is used in cross-country and undersea networks.

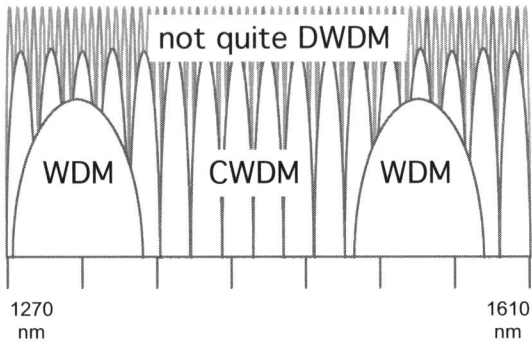

Figure 3-5: Comparison Of WDM, CWDM, And DWDM

Joining and separating multiple wavelengths requires passive devices, the subject of the next chapter.

4 PASSIVE DEVICES

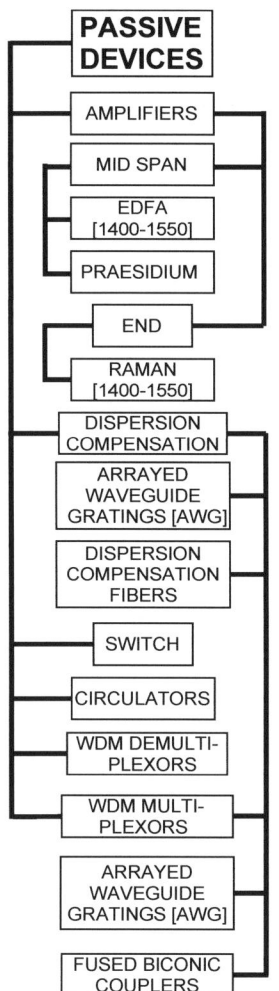

Figure 4-1: Passive Devices

4.1 INTRODUCTION

Passive devices are those that handle or manipulate light as light. The alternative, active devices, are those and handle or manipulate light through conversion of light signals to electrical signals and electrical signals to light signals. Passive devices have three advantages:

- Reduced network cost
- Increased network reliability
- Increased network design flexibility

Conversion of a light signal to electrical signal or an electrical signal to light signal is expensive and can reduce the reliability of the network. Many, but not all, passive devices require no power.

4.2 TYPES

Passive devices include:

- Couplers
- Splitters
- Wavelength division multiplexers
- Wavelength division demultiplexer
- Optical amplifiers
- Dispersion compensators
- Optical switches
- Add drop multiplexers (**ADM**)
- Remotely configurable add-drop multiplexers (**ROADM**)
- Rotary joints

4.2.1 COUPLERS

Couplers combine optical signals, usually, but not always, of different wavelengths (Figure 4-2). Such couplers are actually wavelength division multiplexers. Splitters split signals into multiple output signals, often going to different locations, as in a passive optical network.

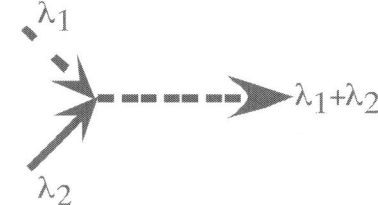

Figure 4-2: Coupling Of Multiple Wavelengths

4.2.2 SPLITTERS

Splitters split the optical power (Figure 4-3). Splitters can have any number of output ports and any splitting ratio. If a splitter splits and separates multiple wavelengths, it is actually a wavelength division demultiplexer (Figure 4-4).

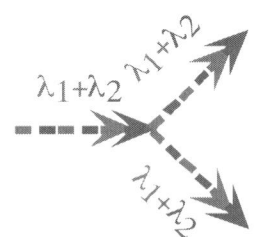

Figure 4-3: Splitting Of Optical Power

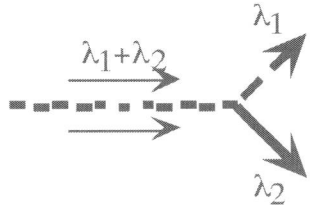

Figure 4-4: Separating Of Multiple Wavelengths

4.2.3 OPTICAL AMPLIFIERS

An optical amplifier is a device that amplifies an incoming optical signal. Such amplifiers can be used in three locations:

- After a transmitter
- Mid span in a link
- Ahead of a receiver

Often, an optical amplifier after a transmitter feeds a splitter to enable a single transmitter to transmit to multiple receiving sites. This configuration compensates for high power loss in the splitter. This compensation enables delivering a power level to the receiver sufficient for accurate recovery of the electrical signal.

There are two types of optical amplifiers: EDFA and Raman. An EDFA amplifier amplifies in an erbium-doped fiber separate from the transmission fiber (Figure 4-5). A Raman amplifier amplifies in the transmission fiber.

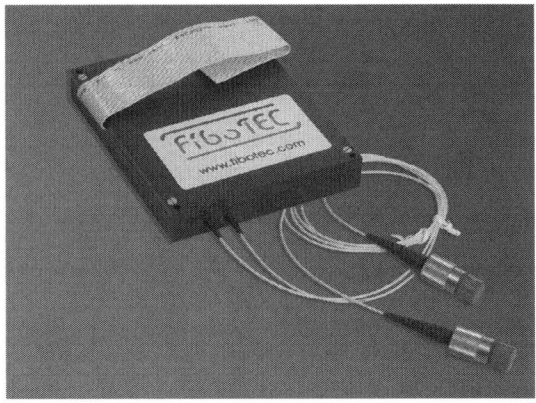

Figure 4-5: EDFA (Courtesy FIBOTEC)

The EDFA uses a length of erbium-doped fiber between two sections of transmission fiber. The Raman amplifier amplifiers the optical signal in the transmission fiber. This difference results in different configurations for these two types of amplifiers. The laser power of the EDFA is confined to the EDFA. The laser power of the Raman amplifier is launched into the communication fiber in the direction opposite to the communication direction. Because optical amplifiers require lasers to provide amplification, opti-cal amplifiers art exception to the general rule that passive devices required no power.

A mid span optical amplifier compensates for power loss in the link. Such loss occurs as attenuation in the fiber, as power lost in connections and in passive devices, such as splitters and wavelength division demultiplexers.

An optical amplifier ahead of an receiver amplifies the optical signal to a level that allows the receiver to accurately recover the electrical signal.

An optical amplifier after transmitter is usually an erbium doped fiber amplifier (EDFA). A mid-span optical amplifier can be an EDFA or a Raman amplifier. An optical amplifier ahead of a receiver can be either type.

Both EDFA and Raman amplifiers function by stimulating atoms in a fiber (Figure 4-6). This stimulation is done by laser light. The electrons in the stimulated fiber are in elevated energy level. Photons in the optical signal strike these atoms, causing the electrons to drop to a reduced energy level. The energy released by this drop as a photon at the same wavelength as that of the photon that disturbed the atom. Thus, the amplifier produces photons of the same wavelength as the signal wavelength.

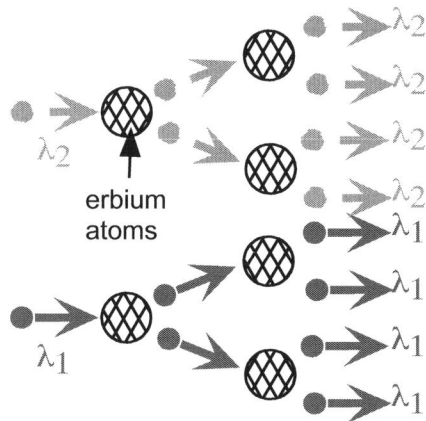

Figure 4-6: Optical Amplification

The EDFA amplifies multiple wavelengths simultaneously. With this property, wavelengths can be added to the fiber without need to reconfigure the EDFA.

4.2.4 DISPERSION COMPENSATORS

Optical amplifiers compensate for power loss and a link. However, dispersion accumulates. To counteract increasing dispersion, fiber networks incorporate dispersion compensating devices. These devices can be dispersion compensating fibers or **arrayed waveguide gratings** (**AWG**). Both of these devices compensate for chromatic dispersion.

Chromatic dispersion, the main form of dispersion in an optical fiber, occurs because transmitters emit a range of wavelengths. This range, the spectral width, results in different wavelengths in the same pulse traveling at different speeds in the fiber. A dispersion-compensating device reverses this chromatic dispersion: wavelengths within the spectral width that travel fast in the fiber, travel slowly in the dispersion compensation device. Wavelengths that travel slowly in the fiber travel fast in the compensating device.

The combination of optical amplification and dispersion compensation results in the ability to transmit to long distances without returning to the electrical signal regime. This author is aware of transmission demonstrations to 7000 km without returning to the electrical or signal regime.

4.2.5 SWITCHES AND ROUTERS

The passive devices so far presented enable transmission of multiple wavelengths to multiple receiving locations and to long transmission distances. However, network planners need to be able to provide communication to locations other than end locations. This need requires other passive components.

Such components are optical switches, add-drop optical multiplexers (**ADM**), and remotely configurable add-drop multiplexers (**ROADM**).

Optical switches enable routing of multiple input signals to multiple output fibers. Such switches use microelectronic mirrors (MEMs) that can be tilted in two axes by piezoelectric effects (Figure 4-7, Figure 4-8 (Figure 4-9)).

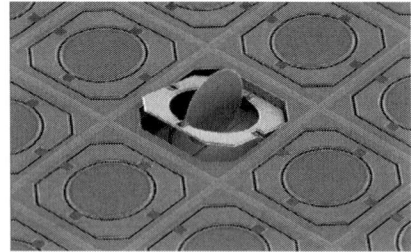

Figure 4-7: Microelectronic Mirrors (www.tf.uni-kiel.de

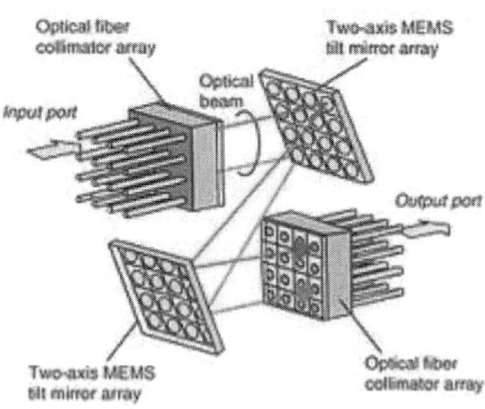

Figure 4-8: Optical Switch Function (opticalengineering.spiedigitallibrary.org)

Add-drop optical multiplexers drop and add specific wavelengths at mid span locations. ROADMs are add-drop multiplexers that can be config-

ured remotely. Of course, none of these massive devices would be useful without the fiber, which is the subject of the next section.

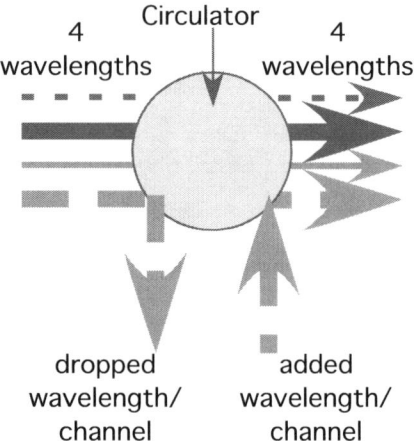

Figure 4-9: Add-Drop Multiplexor

A rotary joint enables movement of the optical signal from a rotating structure to a stationary structure (

Figure 4-10). Submarine sensor cables and cables that control an underwater remotely operated vehicle require this passive device.

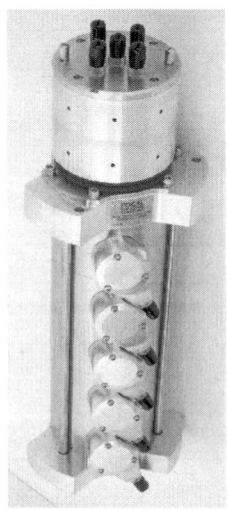

Figure 4-10: Five Fiber Rotary Joint[3]

---

[3] This is a Focal Model 242 from Kaydon Power & Data Technologies. Photograph is courtesy of Kaydon Power & Data Technologies.

5 FIBER

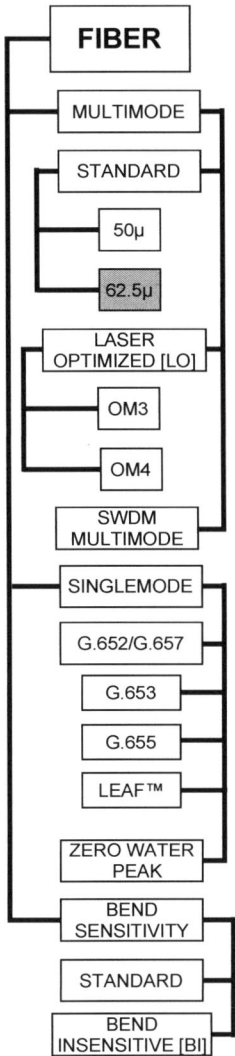

Figure 5-1: Fibers

5.1 FIBER FUNCTION

The function of the fiber is to guide the light between transmitter and the receiver with minimum single distortion. Minimum signal distortion means minimum power loss and minimum dispersion. The fiber fulfills its function through its structure.

5.2 FIBER STRUCTURE

That structure has at least two but usually three regions (). The central region, the **core**, is the region in which most of the optical energy travels. The core can have one of multiple diameters, as presented later. Around the core is the region, the **cladding**, which confines of the light to the core. In addition, the cladding increases the fiber size so that it is easily handled. Glass fibers have a standard cladding diameter of **125 μ**.

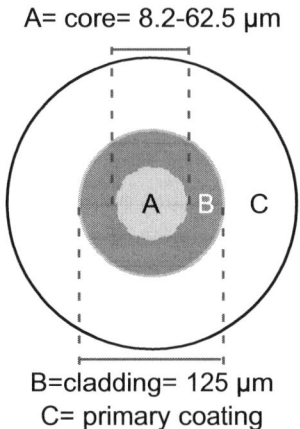

A= core= 8.2-62.5 μm
B=cladding= 125 μm
C= primary coating

Figure 5-2:The Three Regions Of Glass Fiber

Most fibers have a core and cladding made of glass. Because the glass is susceptible to damage from mechanical and chemical attack, the cladding must be protected. This protection comes from a layer call the **primary coating**. This coating has typical a diameter of 245μ.

This primary coating, formerly called the buffer coating, does not, in reality, add strength to the fiber. Rather it protects the cladding surface so the fiber retains its intrinsic high strength.

5.3 FIBER TYPES

As a practical matter, there're two types of fiber: **multimode** and **singlemode**. Multimode fibers have large cores, limited bandwidth, and limited transmission distances. Singlemode fibers have small cores, essentially unlimited bandwidth, and long transmission distances.

5.3.1 MULTIMODE

The core diameter is one of the factors that determine the bandwidth and distance capabilities of the fiber. Multimode fibers have core diameters of 50µ and 62.5µ (Figure 5-3). As the 62.5µ core fiber has performance inferior to that of the 50µ-core fiber, we will ignore it in this document.

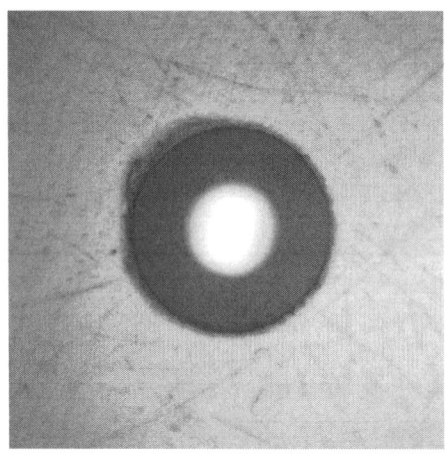

Figure 5-3: Multimode Fiber (~150)

5.3.1.1 LO MULTIMODE FIBERS

The 50µ fibers come in two types: Standard and **laser optimized** (**LO**). These two fibers differ in their maximum transmission distances at 1 Gbps and above. The LO fiber has longer transmission distances at these bit rates than those of a standard fiber. This superior performance results from optimizing the core for use with a VCSEL transmitter. The LO fiber is available at two transmission distances, **OM3** and **OM4**. The OM4 fiber has a transmission distance higher than that of the OM3 fiber.

5.3.1.2 SWDM MULTIMODE FIBERS

A relatively new fiber, the **SWDM** fiber, is designed to enable CWDM transmission over multimode fiber. Before the introduction of this fiber, CWDM was a singlemode transmission technology. This SWDM fiber is being developed to enable transmission of 40 Gbps and 100 Gbps signals over a single multimode fiber pair. At the time of this writing, such transmission requires four and 10 pairs.

5.3.2 SINGLEMODE

The second type of fiber, singlemode, has:

> A very small core diameter (Figure 5-4)
> Essentially unlimited bandwidth
> Very long transmission distances

Singlemode fiber operates at long wavelengths. The combination of small core diameter and long wavelength results in the elimination of the largest form of dispersion, **modal dispersion**. Modal dispersion occurs only in multimode fibers.

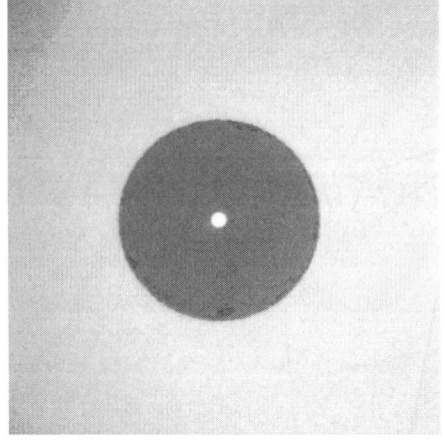

Figure 5-4: Singlemode Fiber (~150)

5.3.2.1 SINGLEMODE TYPES

As is the case with multimode fibers, singlemode fibers come in multiple types. These types are indicated by the designation G.6xx. These designations come from the international telephone and telegraph union.

There are four designations indicating three fiber types. The main difference between the three designnations is the optimum wavelength. The first two designations, **G.652** and **G.657**, have optimum wavelengths of approximately 1310 nm. This fiber type is used in telephone networks, CWDM networks, and FTTH networks.

The third designation, **G.653**, has an optimum wavelength 1550 nm. This fiber type has the advantages of both low attenuation rate and low dispersion. At the time of this writing, this author's impression is that this fiber is rarely used.

The fourth designation **G.655**, has an optimal wavelength close to, but not at, 1550 nm. This fiber type was designed specifically for DWDM networks. A unique form this fiber type is **LEAF**™ fiber (Corning Inc.). The LEAF G.655 fiber is able to transmit power levels higher than those possible with the Standard G.655 fiber.

Singlemode fibers may or may not have a water peak (Figure 5-5). A water peak is an increase in attenuation rate that occurs at approximately 1380 nm. Such a peak makes transmission of CWDM or DWDM signals at that wavelength difficult. Without such a peak, fiber is known as a '**zero water**' fiber or '**no water**' fiber.

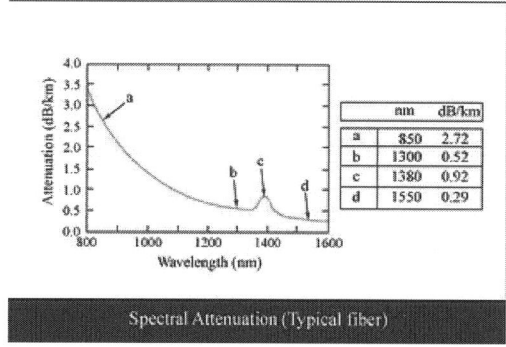

Figure 5-5: Attenuation Rate With Water Peak (MIT Open Courseware)

5.3.3 BEND SENSITIVITY

Both multimode and singlemode fibers are available with two levels of bend sensitivity: Standard and bend insensitive (BI). Bend insensitive fibers can be bent to radii smaller than those of standard fibers with power loss increases less than those of standard fibers.

Bend insensitive fibers cannot be used in all applications. Multimode BI fibers cannot be used in test leads for insertion loss testing of some multimode fibers.

Of course, the glass fiber with an overall diameter of 250μ cannot be used as is. The fiber must be protected against damage during installation and use. Such protection comes from the cable, the subject of the next chapter.

6 CABLES

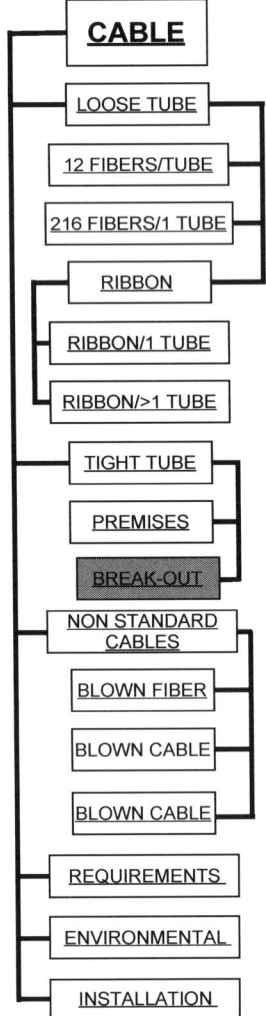

Figure 6-1: Cables

6.1 INTRODUCTION & STRUCTURE

Fiber optic cable is a package that protects the fiber from damage during installation and use. This protection comes from the structural elements. These elements include:

- Buffer tubes
- Strength members
- Water blocking materials
- Jackets
- Steel armor

The **buffer tube** is the first layer of plastic placed around the fiber by the cable manufacturer. The type of buffer tube determines the cable design. There are two types of buffer tubes: **loose buffer tube** and **tight buffer tube**.

In a loose buffer tube cable, the fibers are encased in an oversized tube (Figure 6-2). As a result, the buffer tube can enclose multiple fibers. Common fiber counts are 12 and 216.

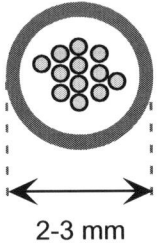

Figure 6-2: Loose Tube

In a tight buffer tube cable, the inside diameter of the buffer tube is exactly the same as the outside diameter of the fiber (Figure 6-3). As a result, a tight buffer tube has one fiber per buffer tube.

Figure 6-3: Tight Tube

6.2 CABLE DESIGNS

6.2.1 4 LOOSE TUBE DESIGNS

Four loose tube cable designs are common. The first design is a multiple fiber for tube design (this author's designation, Figure 6-4).

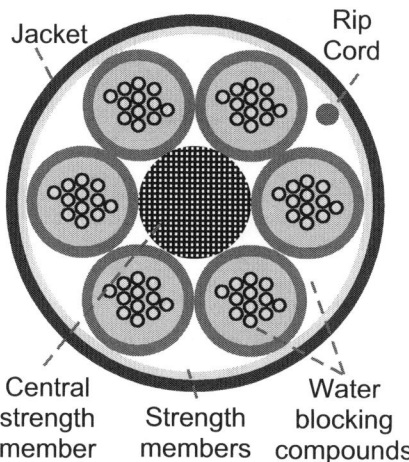

Figure 6-4: MFPT Design

The second design has a single large loose buffer tube in the center of the cable (Figure 6-5). This loose buffer tube contains multiple bundles of fiber. The bundles contain 12 color-coded fibers and are held together with color-coded threads.

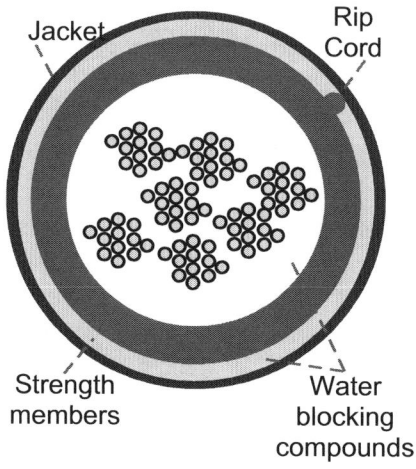

Figure 6-5: Central Loose Tube Design

The third design is a modification of the second design. This third design has a single large loose buffer tube in the center of the cable. However, in this design, the fibers are arranged in stacked **ribbons** (Figure 6-5, Figure 6-6). A ribbon has 4 to 24 fibers precisely aligned and glued to a tape substrate.

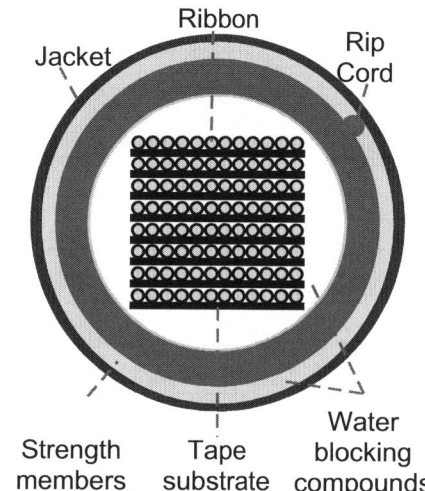

Figure 6-6: Ribbon Cable Structure

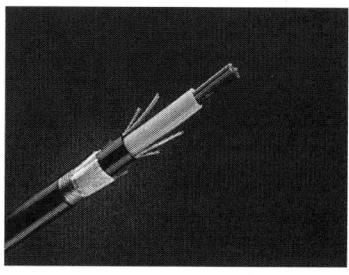

Figure 6-7: Ribbon Cable (Courtesy Corning Inc.)

The fourth design combined features of the first and third designs (Figure 6-8). In this design, multiple ribbons are placed in multiple barber tubes. The buffer tubes are stranded around a central strength member.

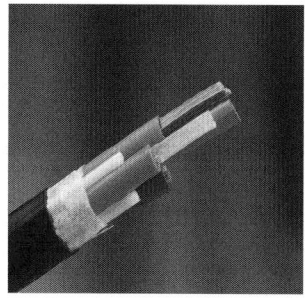

Figure 6-8: High Count Ribbon Cable

6.2.2 2 TIGHT TUBE DESIGNS

Two tight tube cable designs are common. Both are known as **premises**, or **distribution**, cables (Figure 6-9). The basic premises design consists of up to 24 fibers in a single jacket. The second comment design repeats the first design as a sub cable (Figure 6-10).

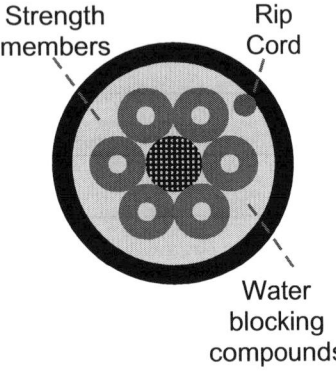

Figure 6-9: Premises Cable Structure

Figure 6-10: High Count Premises Cable Structure

6.3 PERFORMANCE REQUIREMENTS

The performance requirements that the cable must meet depend upon the installation and environmental conditions to which the cable is exposed. Such conditions include, but are not limited to:

- Moisture resistance
- Crush load resistance
- Bend radius limitation, short-term
- Bend radius limitation, long-term
- Temperature operating range
- Temperature storage range
- Temperature installation range
- Long-term use load
- Chemical resistance
- Insulation load
- National electrical code (NEC) compliance

NEC compliance requires use of cables that meet requirements of the location in which the cables are installed. The NEC code identifies eight cable ratings. Two of these ratings, **OFNR** and **OFNP**, are commonly used. OFNR rated cable is used in horizontal and riser locations. OFNP rated cable is used in air handling platinum locations.

6.4 COLOR CODES

Cable jackets and fibers have color-coding. Three cable jacket colors are recognized by the standards:

- Yellow
- Aqua
- Orange

Singlemode cables are **yellow**. LO cables are **aqua**. Multimode cables are **orange**. In addition, outdoor cables tend to have a black jacket.

Fibers and buffer tubes have a **color code sequence**:

1. Blue
2. Orange
3. Green
4. Brown
5. Slate
6. White
7. Red
8. Black
9. Yellow
10. Violet
11. Rose
12. Aqua

This sequence enables cables to contain up to 144 fibers. For fiber counts above 144, a dash or stripe is added to the buffer tube. For fiber counts about 288, multiple dashes or stripe are used.

7 CONNECTORS

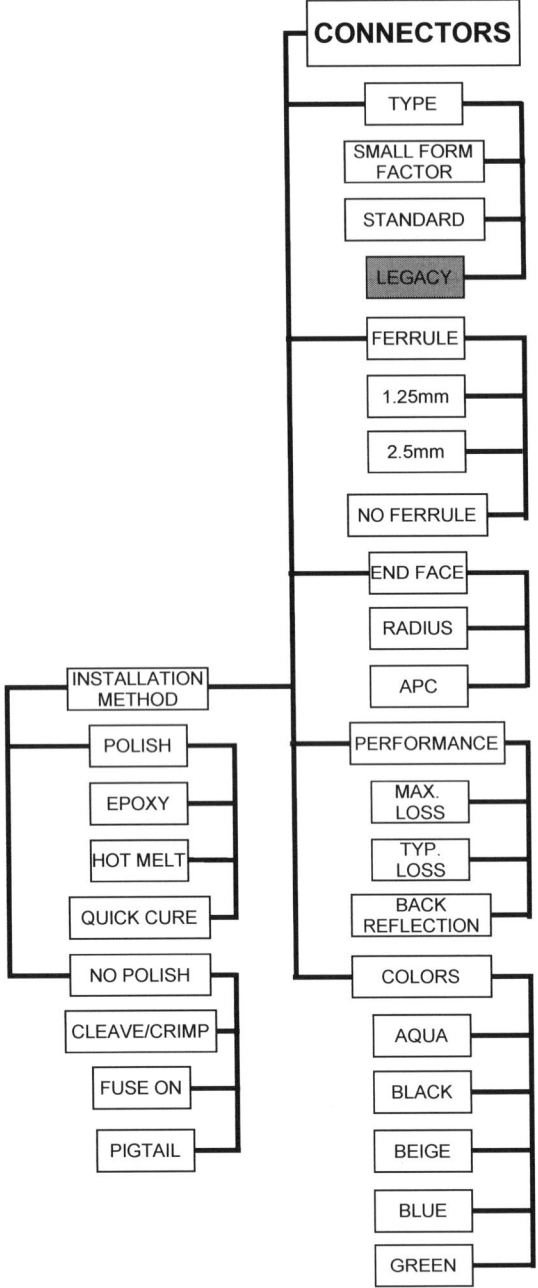

Figure 7-1: Connectors

7.1 FUNCTIONS

Fiber optic connectors have multiple functions. The first, and primary, function is to enable disconnection and rerouting of the signal. Other functions include: low power loss, low back reflection, and end face protection.

All but one of the current generation of connectors have one feature in common: the ferrules make physical contact (Figure 7-2). Contacting ferrules result in reduced power loss and reflectance, aka 'back reflectance'.

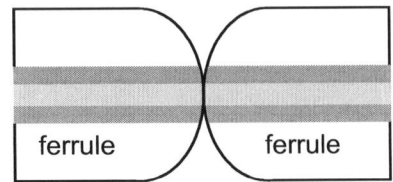

Figure 7-2: Contact Ferrules With Radius

In connectors with a ferrule, such contact requires a radius of curvature on the tip of the ferrule. This radius eliminates the need to create perfectly smooth, perfectly perpendicular end faces.

7.2 CONNECTOR TYPES

Fiber optic connectors are of three types:

- Small form factor (SFF)
- Standard
- Legacy

The small form factor connectors are those most recently developed. Most became available after approximately 1997. The small form factor enables increased port count at patch panels and in switches. SFF connectors include:

- LC
- LX.5
- Volition
- Opti Jack
- MTRJ
- MU
- MTP/MTO

The Volition (Figure 7-3), MTRJ, and Opti-Jack are duplex connectors. The MTRJ connector is experiencing reduced demand and support. The Volition connector is unique in that it is the only fiber optic connector that has no ferrule.

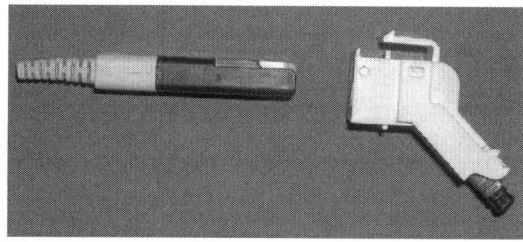

Figure 7-3: Volition Plug And Jack

The LC (Figure 7-4), LX.5, and MU are simplex connectors that can be converted to duplex with hey external clip (Figure 7-5). The MTP/MTO connector (Figure 7-6) is slightly larger and the SC connector (Figure 7-8). The MTP/ MTO connector is avail-able in versions for 12 fibers, 24 fibers, and 72 fibers.

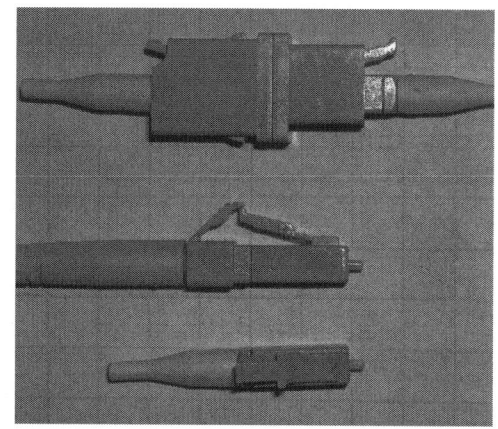

Figure 7-4: LC Connector

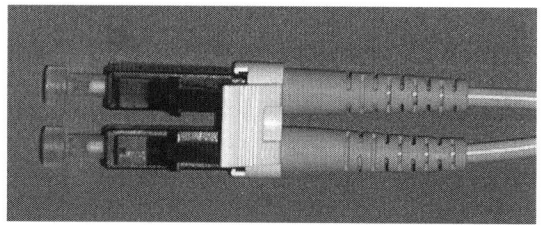

Figure 7-5: LC Duplex Connector

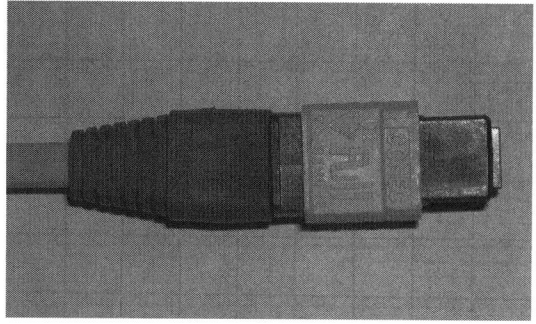

Figure 7-6: MTP/MTO Connector

All of these connectors are roughly half the size of the prior Standard connectors, the SC and ST compatible.

The ST®-compatible (Figure 7-7) and the SC connectors (Figure 7-8) became available in the mid 1980s and have seen extensive use since that time. Although the SC connector has performance superior to that of ST-compatible connector, it is in common use today.

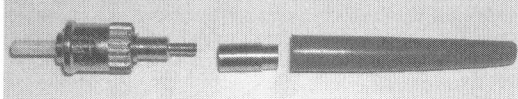

Figure 7-7: ST™-compatible Connector

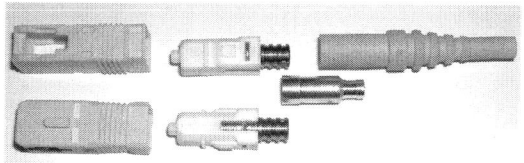

Figure 7-8: SC Connector

The standard and small form factor connectors have three common features. All make physical contact. That is there is no deliberate air gap between two connectors at a patch panel. Most of the legacy connectors that preceded these two types had a deliberate air gap between mated connectors.

The second common feature is the existence of the key. This key enables consistent power loss from insertion to insertion by preventing rotation of the ferrule relative to the mated ferrule at a patch panel. Most legacy connectors had no key.

The third common feature is the value of the maximum loss. That value is always **0.75 dB/pair**.

Two ferrule diameters dominate the industry: **2.5 mm** and **1.25 mm**. The **ST™-compatible**, **SC**, and the Opti-Jack have 2.5 mm diameter ferrules. The **LC**, **LX.5**, and **MU** have a 1.25 mm diameter ferrule. This small diameter ferrule has power loss lower than that of the 2.5 mm diameter ferrule. For example, 2.5 mm diameter ferrule has a typical loss of 0.3 dB/pair. The 1.25 mm diameter ferrule has a typical loss between 0.15 and 0.2 dB/pair.

The ferrules can have one of 3 faces: **radiused** (Figure 7-2), **angled** (**APC**) (Figure 7-9), or flat. Small form factor and standard connectors have radiused end faces. A radius end face is necessary to achieve the low level of back reflection required in today's high bandwidth networks. Excessive back reflection in high-bandwidth networks can create transmission errors, even if Power loss is low and dispersion is sufficiently low. SC, LC, FC, and LX.5 connectors are available with an angled (APC) end face.

Most of the legacy connectors have flat end faces. Flat ferrule end faces produce high levels of back reflection. Because of this performance disadvantage, Flat end faces are not acceptable for use in high bandwidth networks.

7.3 COLOR CODING

Connectors, their boots, or the barrel in the patch panel need to be color-coded in order to indicate the type of fiber the backside of the patch panel or in the patch cord. The Building Wiring Standard, EIA/TIA–568–C [4] indicates use of five colors:

- Aqua
- Black
- Beige
- Blue
- Green

Aqua connectors contain OM3, or OM4 fiber. **Black** connectors contain standard 50μ core diameter fiber, which is designated OM2. **Beige** connectors contain the 62.5 μ fiber, which is designated **OM1**. **Blue** connectors contain singlemode fiber. **Green** connectors contain single mode fiber and have and an APC end face.

Figure 7-9: SC/APC Connectors

Connectors can be installed by at least seven different methods, which can be organized into two distinct groups. The first group of installation methods requires polishing. The second group does not.

---

[4] Note: TIA/EIA-568-D is in review and is expected to be issued in the near future (as of 9/25/15).

7.4 INSTALLATION METHODS

7.4.1 POLISH METHODS

At least three commonly used connector installation methods require polishing. They are:

- Epoxy
- Hot melt adhesive
- Quick cure adhesive

The epoxy method allows installation of the lowest-cost connectors. However, the time for installation is highest.

The hot melt adhesive method requires use of connectors that are preloaded with hot melt adhesive. Such connectors have costs higher than those of epoxy connectors. However Time for insulation is lower than that for epoxy connectors.

The quick cure adhesive is an anaerobic-like adhesive. It is actually an edge filling adhesive, which hardens when in a thin film, like that between the outside of the fiber and the inside of the fiber hole in the ferrule. It is usually two part, an adhesive and an accelerator. The quick cure adhesive method allows use of low cost connectors. The installation time is about the same as that for the hot melt adhesive method. This author's experience indicates that process yield with quick cure adhesives by novice installers will be the lowest of these three methods.

7.4.2 NO POLISH METHODS

Three connector installation methods require no polishing:

- cleave and crimp
- Fuse on
- Pigtail slicing

All three of it these methods required connectors with fibers installed in the factory. These three methods differ in the amount of fiber in the connector.

The 'cleave and crimp' connector installation method, was pioneered by Corning Incorporated in the mid 1990s. This method involves use of a connector with a preinstalled fiber (Figure 7-10). The fiber end in the ferrule is polished a factory. The fiber end inside the connector is cleaved. That is, the fiber is broken in a manner to create an internal fiber end that is nearly perfectly perpendicular and nearly perfectly smooth. In essence, this connector is a mechanical splice in the connector back shell.

This method requires a cleaver and a unique installation tool (Figure 7-11). The installation tool provides indication of low loss prior to completing the installation. Use of this tool has enabled Corning to offer a guarantee of 100% yield, through replacement of high loss connectors. This author's experience with this product justifies this guarantee. The prime advantage of this tool is fast installation. This author's web searches revealed claims of 30/hour, 40/hour and 60/hour.

This method is preferred in four situations. The first situation is the organization that has high personnel turnover, such as military and governmental groups. The second situation is installation in unfavorable environments. Such environments include: deserts, dirty air, bucket trucks, extremes of environment, and locations requiring few connectors.

The third situation is that of extremely high total loaded labor rates. The fourth situation is that of infrequent installations.

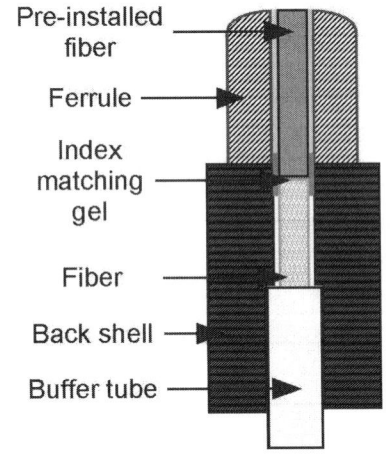

Figure 7-10: Structure Of Unicam® Connector

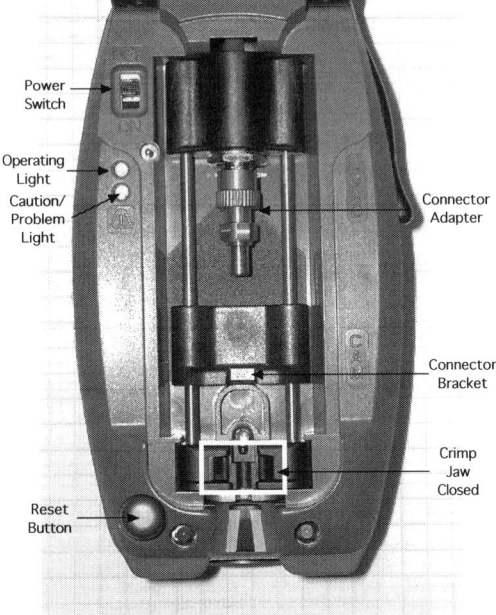

Figure 7-11: Unicam® Installation Tool

A modification of the cleave and crimp connector is the fuse-on connector. Like the Unicam®, this connector has a pre-installed fiber. The ferrule end of the fiber is polished in the factory. The other end of the fiber extends out of the back shell of the connector.

This method requires use of fusion splicer that has a holder compatible with the connector. The fusion splicer splices of fiber in the cable to the fiber in the connector. A connector installed by this method has one of the highest costs of all the connectors. However, installation time is one of the lowest of all the methods.

The third connector installation method that requires no polishing is fusion splicing of a pigtail to the cable. A pigtail is a short length of cable or 900 μ tight buffer tube with a factory-installed connector on one end.

In this method, a pigtail with a diameter between 900μ and 3 mm is spliced to the cable. The splice is contained in a splice tray, an item of cost that is not necessary for Cleave and Crimp or fuse-on connectors.

This brief presentation clearly indicates that choice of connector installation method is a somewhat complex process. In different situations, the lowest total installed cost will be different.

8 SPLICES

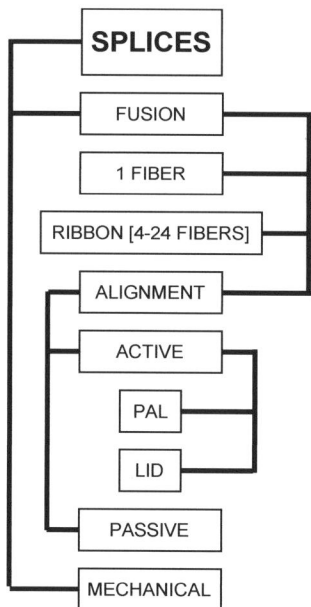

Figure 8-1: Splicing

8.1 INTRODUCTION

Splicing is the permanent or semi-permanent connection of two fibers. Splicing is performed under at least three conditions. Such conditions include:

- Inability to purchase the required length as a single length
- Inability to install a cable and a single length
- Repair of a broken cable (also known as backhoe fade, drunk driver fade, etc.)

8.2 TWO METHODS

Two methods of splicing can be performed: **fusion splicing** and **mechanical splicing**. Fusion splicing, the preferred method in most cases, is done by melting together two pieces of glass fiber. One could call this glass welding. The fusion splicer (Figure 8-2) provides alignment of the fiber cores of singlemode fiber. The fusion splicer provides alignment of the claddings of multimode fibers.

Figure 8-2: Typical Fusion Splicer

8.3 SPLICING STEPS

Splicing requires the following steps:

- installation of a heat shrinkable splice cover
- removal the primary coating
- cleaning the fiber
- cleaving the fiber
- alignment fibers and the fusion slicer
- fusing the fibers
- Heating the splice cover

The cleaving produces a nearly perfectly perpendicular and nearly perfectly smooth end face on the fiber. Such an end face is necessary to achieve low-power loss.

The **splice cover** (Figure 8-3) supports the fiber and isolates the fiber from the environment. Alternatively, an **adhesive splice cover** is placed on the fiber after of the splice has been made.

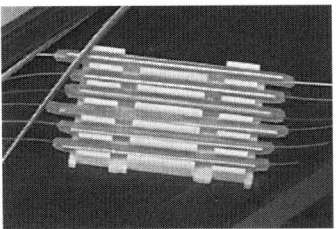

Figure 8-3: Splice Covers In Holder

Mechanical splicing is splicing is the connection of two fibers with a mechanical device. The device provides alignment of the two fibers. Such alignment is necessary to achieve low power loss. The mechanical slice isolates the fiber from the environment.

After making the slice, the installer places the fiber in a **splice tray** (Figure 8-4). After all splices are in the tray, the installer places the tray in an **enclosure** (Figure 8-5).

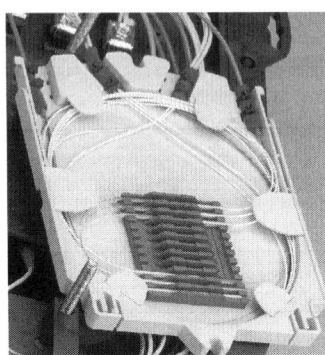

Figure 8-4: Splice Tray With Pigtails

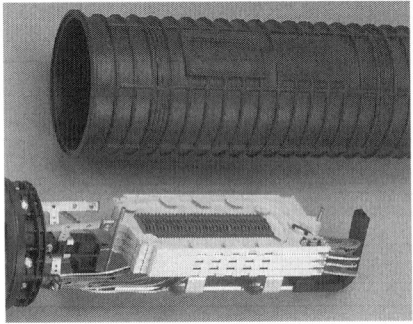

Figure 8-5: Typical Outdoor Enclosure

9 TESTING

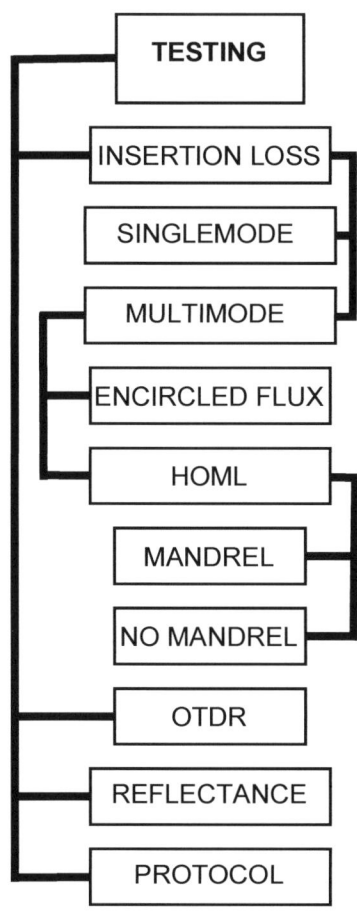

Figure 9-1: Testing

9.1 TYPES

Fiber optic links can undergo four types of testing:

- Insertion loss
- OTDR
- Back reflection
- Protocol

9.2 INSERTION LOSS TESTING

The first type of testing, insertion loss testing is always required by the Building Wiring Standard, TIA/EIA-568-C. This test comes close to simulating a loss that a transmitter receiver pair will experience. To achieve this simulation, the test is performed at the same wavelength as that of the transmitter. This test is performed by measuring an input power level between a light source and a power meter (Figure 9-2). The output power is measured when the light source is connected to one end of the link and the power meter to the other end (Figure 9-3).

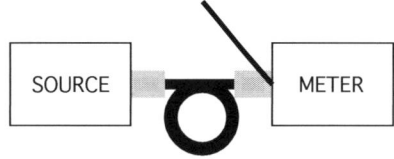

Figure 9-2: Measurement Of Input Power, Singlemode

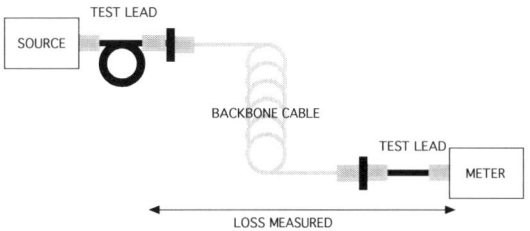

Figure 9-3: Measurement Of Output Power, Singlemode

While simple in concept, the insertion loss test is somewhat complex in practice. Testing a single-mode and multimode links requires different procedures. In addition, testing of 50μ links at 850 nm requires procedures different from testing 62.5μ fibers and 50μ fibers at 1300 nm. Finally, 50μ multimode link testing at 850 nm can be performed with at least two types of equipment.

We present a summary of single-mode and multimode link testing below. The details of this testing appear in Chapter 14 of Professional Fiber Optic Installation, v.9.

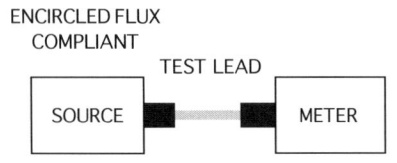

Figure 9-4: Measurement Of Input Power, 50μ @ 850 nm

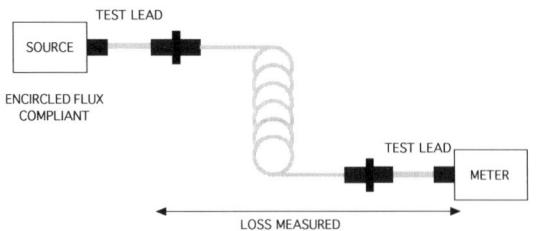

Figure 9-5: Measurement Of Output Power, 50μ @ 850 nm

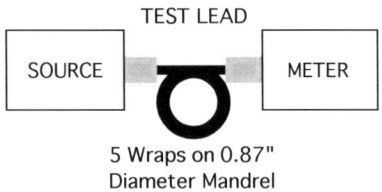

Figure 9-6: Measurement Of Input Power, 62.5μ @ 850 nm and 1300 nm

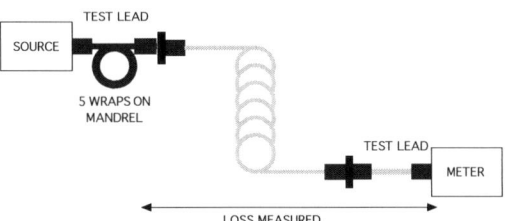

Figure 9-7: Measurement Of Output Power, 62.5μ @850 nm and 1300 nm

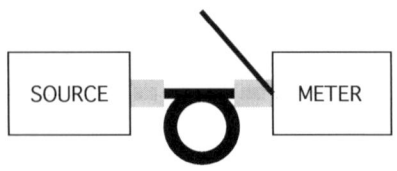

Figure 9-8: Measurement Of Input Power, 50μ @ 1300 nm, HOML

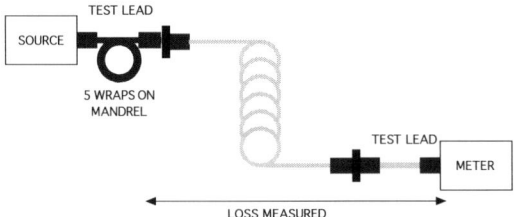

Figure 9-9: Measurement Of Output Power, 50μ @ 1300 nm

9.3 OTDR TESTING

The second type of testing is OTDR testing. The Building Wiring Standard, TIA/EIA-568-C does not require optical time domain reflectometry (OTDR) testing. However, this author, and many other professionals, recommend such testing. Such testing can indicate hidden conditions of reduced reliability.

OTDR testing requires an OTDR (Figure 9-10, Figure 9-11). The OTDR creates a trace (Figure 9-12, Figure 9-14) that indicates cable segments, splice locations, connector locations, and locations at which there is a stress on the cable.

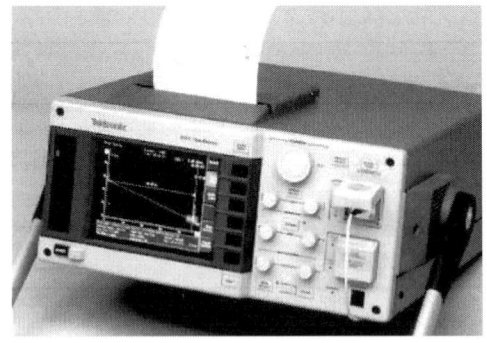

Figure 9-10: Mainframe OTDR

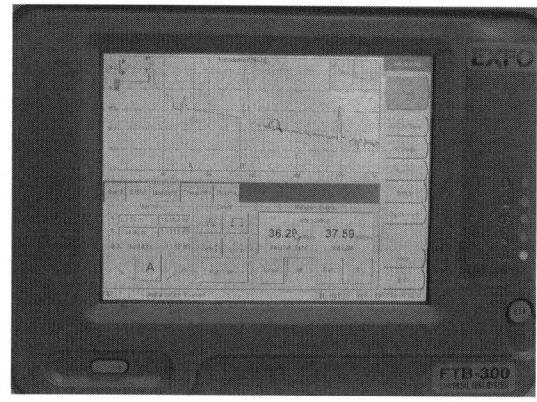

Figure 9-11: Mini-OTDR

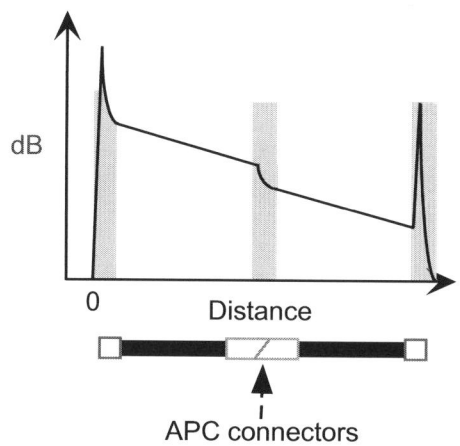

Figure 9-12: Generic Two Segment OTDR Trace

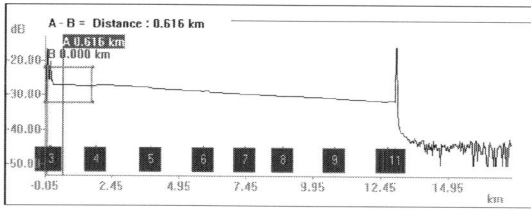

Figure 9-13: 17 Km, 1310 nm Singlemode Trace

Interpretation of a trace requires an accurate map of the link and knowledge how to place cursors on the trace to make accurate measurements (Figure 9-14, Figure 9-15). This knowledge is detailed in Chapter 16 of Professional Fiber Optic Installation, v.9 and in Mastering The OTDR-Trace Acquisition And Analysis.

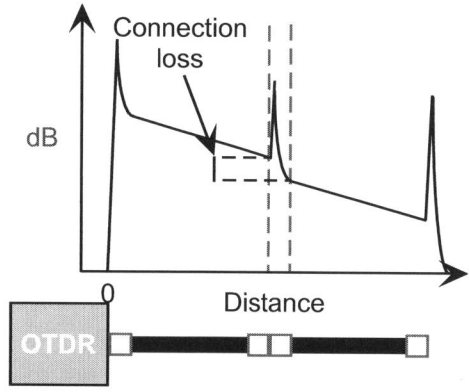

Figure 9-14: Curser Placement For Connection Loss Measurement

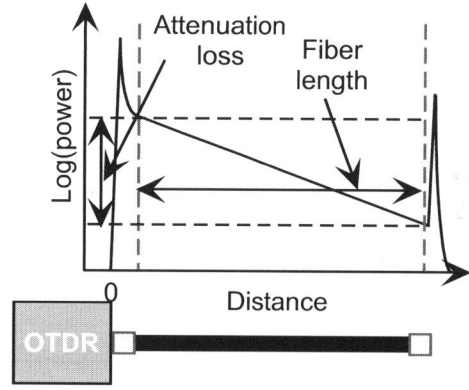

Figure 9-15: Curser Placement For Attenuation Rate Measurement

9.4 REFLECTANCE TESTING

The third type of testing is back reflection testing. Back reflection testing requires a back reflection test set (Figure 9-16). Back reflection testing indicates the power level reflected back from connectors and other passive devices (Figure 9-17, Figure 9-18). Such reflected power can corrupt the optical pulse stream. Such corruption can result in long connect times, low throughput, and inability to communicate over the link.

9.5 PROTOCOL TESTING

The fourth type of testing is protocol testing. This testing is done on the electrical signal output from the receiver. Such testing can indicate problems with the electrical signal. Such testing can differentiate between optical problems and electrical signal problems.

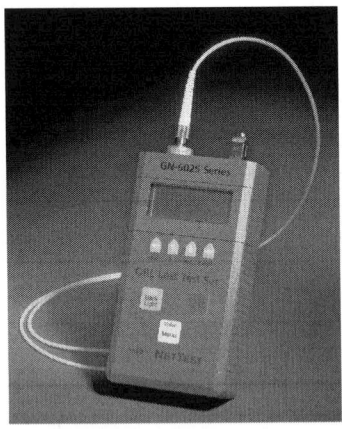

Figure 9-16: Reflectance Test Set (Courtesy of NetTest)

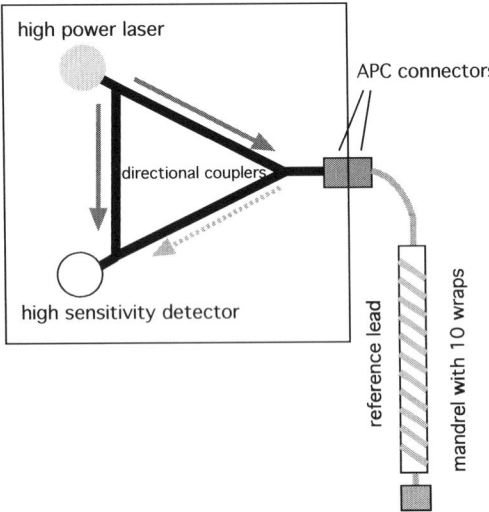

Figure 9-17: Reflectance Calibration

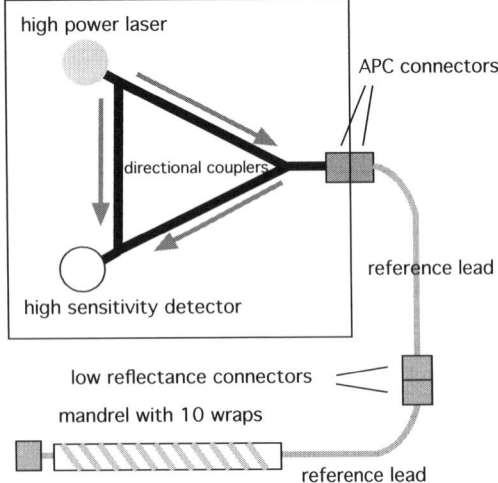

Figure 9-18: Reflectance Test

10 DESIGN CONCERNS

The fiber optic network designer has six objectives. These objectives are:

1. Achieving sufficient power at the receiver
2. Minimizing power loss through link
3. Avoiding excessive power at the receiver
4. Achieving sufficiently low dispersion
5. Achieving high reliability of all link components during installation and use
6. Minimizing cost

The first two objectives require choosing link components which, when assembled into a link, have an end-to-end power loss that is less than the maximum loss, known as the **optical power budget (OPB)**, The OPB is the maximum power loss at which the optoelectronics will function with the required level of accuracy.

The third objective requires knowledge of the minimum loss requirement of the optoelectronics. In general, multimode optoelectronics that are compliant with the relevant standard (e.g., 10 Gb Ethernet, Fiber Channel) cannot overload the receiver.

Compliant singlemode optoelectronics, can overload the receiver. Some singlemode optoelectronics are designed for extended transmission distance. Such optoelectronics can overload their receivers when used on short links. In such cases, the designer specifies use of an **attenuator**. This author's experience is that attenuators are placed at the receiver end of a fiber.

The fourth objective requires optoelectronics suitable for the transmission distance. Most standards include combinations of characteristics to enable cost-effective combinations of components to meet distance and bandwidth requirements. These characteristics are:
- fiber core diameter
- wavelength
- fiber dispersion specifications

In general, multimode links can transmit 10 Gbps to short distances. Short distances range from 220 m (722 feet) to 1000 m (3280 feet). Combinations of core diameters, dispersion specifications (i.e., OM2, OM3, OM4), and wavelengths enable matching the required transmission distance to a cost-effective combination.

In general, singlemode fibers are used at 10 Gbps for distances above the capabilities of multimode fiber. Such distances range from 1000 m to 70 km, depending on wavelength and singlemode fiber specifications.

The fifth objective requires knowledge of installation and environmental conditions. Such conditions include, but are not limited to:
- installation load
- installation bend radius
- temperature installation range
- temperature storage range
- temperature operating range
- long term bend radius
- chemical resistance
- rodent resistance
- crush load, long term
- crush load, short term

Such knowledge enables creation of performance specifications for all components. Such specifications result in high reliability.

The sixth objective requires developing knowledge of the product and labor costs. Some low cost products have high labor costs. Some high cost products have low labor costs.

As an example, consider the trade-off between connector costs and labor costs. Low cost connectors are epoxy connectors. Epoxy connectors require high man-hours to install.

In low labor cost situations, low cost connectors will produce the lowest total installed cost. In high labor cost situations, the same low cost connectors will produce high total installed cost. In these high labor cost situations, use of an increased cost connector that has reduced labor cost may produce the lowest total installed cost.

11 ABOUT THE AUTHOR

The author, Eric R Pearson, has 38 years of experience in fiber optic communications. He has managed fiber and cable manufacturing plants. He has designed and developed manufacturing processes for more than 200 cable products. He has developed connector installation procedures. He founded and has managed Pearson Technologies since 1980.

He has made more than 530 fiber optic presentations. He has trained more than 8800 people in fiber optic network design and installation programs that he developed.

For 12 years, he was a founding Director of the Fiber Optic Association (FOA). His duties included development of the certification process, advanced certification requirements, and certification examinations.

The FOA has designated Mr. Pearson a 'Master Instructor'. He has received the following FOA advanced certification designations:

- CFOS/C: Connector Installation
- CFOS/S: Splicing
- CFOS/T: Testing
- CFOS/I: Instructing

For 13 years, he was a Master Instructor for BICSI. He developed and delivered FO110, a fiber optic design course.

For 18 years, he was an Editorial Advisor to Fiberoptic Product News.

Mr. Pearson has provided expert witness support to attorneys in 18 legal cases involving fiber optic products, performance, system damage, and patent issues.

He has written the following books:

- Professional Fiber Optic Installation, v.9 -The Essentials
- Mastering Fiber Optic Network Design
- Mastering The OTDR-Trace Acquisition And Analysis
- Mastering Fiber Optic Connector Installation
- Successful Fiber Optic Cable Installation- A Rapid Start Guide
- Fiber Optical Communications For Beginners- The Basics
- The Complete Guide To Fiber Optic Cable System Installation

Single copies of these texts are available from Amazon.com. Multiple copies of these texts for training are available from Pearson Technologies Inc.

For training purposes, PowerPoint slide books for these programs are available:

- Professional Fiber Optic Installation, v.9
- Mastering Fiber Optic Network Design
- Mastering Fiber Optic Network Design- Management Presentation Slides
- Fiber Cable Testing, Certification And Troubleshooting

Single copies of these texts are available from Amazon.com. Multiple copies of these texts for training are available from Pearson Technologies Inc.

Mr. Pearson delivers the following training programs worldwide:

Professional Fiber Optic Installation, v.9, 4 days

Professional Fiber Optic Installation, v.9 with FOA CFOT certification, 5 days

Advanced Professional Fiber Optic Installation, v.9, 5 days

Advanced Professional Fiber Optic Installation, v.9 with FOA CFOT certification, 6 days

Advanced Professional Fiber Optic Installation, v.9 with FOA CFOT and CFOS/C certification, 6.5 days

Basic Fiber Optic Installation, 3 days

Fiber Cable Testing, Certification And Troubleshooting, 4 days

Eric R Pearson

Mr. Pearson has degrees from MIT (BS) and CWRU (MS), both in Metallurgy and Materials Science.

Contact Information

Eric R. Pearson, CFOS/T/C/S/I
Pearson Technologies Inc.
4671 Hickory Bend Drive
Acworth, GA 30102-6340
770-490-9991
www.ptnowire.com
fiberguru@ptnowire.com

36 Years Of Superior Fiber Optic Training And Consulting

25757720R00024

Printed in Great Britain
by Amazon